HAVE A NICE DNA

Fran Balkwill & Mic Rolph

Cold Spring Harbor Laboratory Press

Development Manager	Jan Argentine
Production Manager	Denise Weiss
Desktop Editor	Danny de Bruin
Project Coordinator	Maryliz Dickerson
Production Editor	Mala Mazzullo

Library of Congress Cataloging-in-Publication Data

Balkwill, Frances R.
Have a nice DNA / Fran Balkwill & Mic Rolph.
p. cm. – (Enjoy your cells ; 3)
Summary: Explains the structure and function of DNA.
ISBN 0-87969-614-1 (alk. paper) – ISBN 0-87969-610-9 (pbk. : alk. paper)
1. DNA–Juvenile literature. (1. DNA.) I. Rolph, Mic. II. Title.
III. Series.
QP624 .B354 2001
611'.01816–dc21
2001042137

10 9 8 7 6 5 4 3 2 1

All CSHL Press publications may be ordered directly from Cold Spring Harbor
Laboratory Press, 500 Sunnyside Boulevard, Woodbury, New York 11797-2924.
Phone: 1-800-843-4388 (Continental U.S. and Canada). All other locations (516)
422-4100. FAX: (516) 422-4097. E-mail: cshpress@cshl.org. For a complete catalog
of all Cold Spring Harbor Press publications, visit our World Wide Web Site
http://www.cshlpress.com

Page 31: This view of the Earth seen by the Apollo 17 crew traveling toward the
moon was supplied courtesy of NASA and the NSSDC.

(Please note that most cells are gray and grainy. We have added some false
color to make the cells in this book look exciting!)

Once upon a time, you were very very small. In fact, you were made of just one tiny cell. But the incredible thing about that tiny cell was that all the instructions to make you were hidden inside it. And all because of a very important chemical substance called DeoxyriboNucleic Acid (Dee-ahk-see-rye-boh-New-clay-ik Acid) —everyone calls it **DNA**.

You started life as one single cell but now you are made of about one hundred million million of them. Your cells are very small, most of them no more than 1/1000 inch wide.

And the amazing fact is that the same DNA from your first cell is in almost every cell in your body.

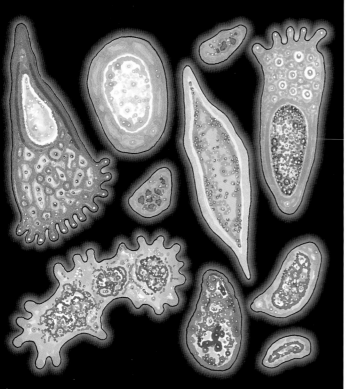

What exactly is DNA? How does it work? Let's look inside one of those cells and find out.

This is a cell. On the outside there is a membrane, a bit like a soap bubble. This membrane encloses jelly-like cytoplasm (sigh-toe-plaz-um) which is crammed with many blob-like objects and tubes. There is also a nucleus (new-clee-us) that controls the cell.

The strange-looking bits inside the cell do all kinds of clever jobs; mitochondria (my-toe-con-dree-ah) turn food into energy and ribosomes (rye-boh-sohms) make important chemicals called proteins (pro-teens).

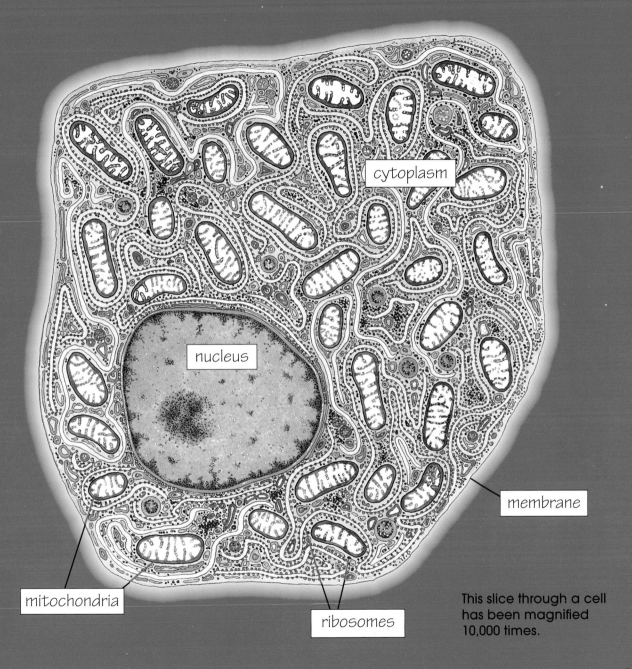

cytoplasm

nucleus

membrane

mitochondria

ribosomes

This slice through a cell has been magnified 10,000 times.

Humans have 23 pairs of chromosomes.

The membrane around the nucleus has disappeared, and dark shapes have appeared in the middle of the cell. The shapes are called chromosomes (kro-muh-sohmz). Chromosomes are made of that amazing chemical substance, DNA.

Each of your cells has forty-six chromosomes. And each chromosome is one long piece of DNA.

You cannot always see the DNA in a cell because it is wound quite loosely inside the chromosomes. But in this cell, the DNA has become tightly coiled. This is because the cell is going to divide into two cells.

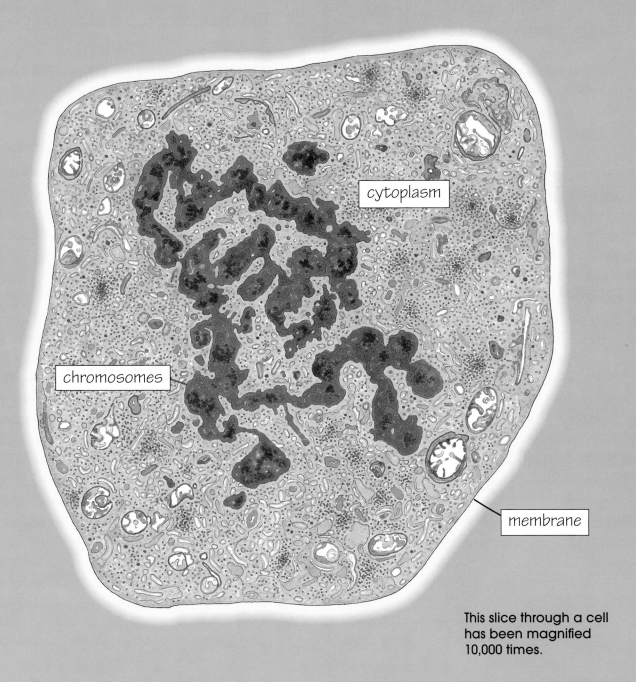

cytoplasm

chromosomes

membrane

This slice through a cell
has been magnified
10,000 times.

The DNA unraveled here has been magnified twenty million times. At last you can see your DNA and discover its secret.

You will observe that DNA is made of not one, but two strands. Each strand is coiled in a shape called a helix.

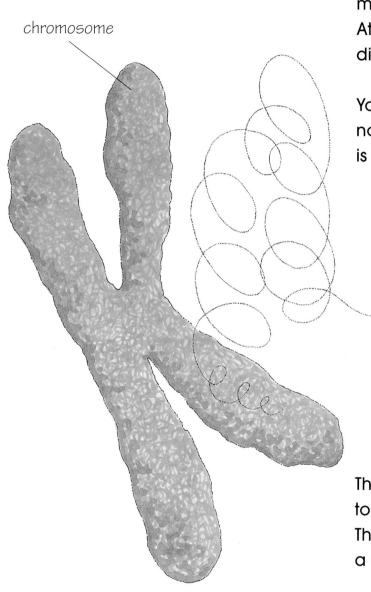

chromosome

The two strands are connected together in a very special way. They make a twisting, turning spiral— a double helix.

Each strand of DNA is made of just four chemicals.

A is a chemical called **A**denine (Ad-en-een)
T is a chemical called **T**hymine (Thigh-meen)
C is a chemical called **C**ytosine (Sigh-toe-seen)
G is a chemical called **G**uanine (Gwa-neen)

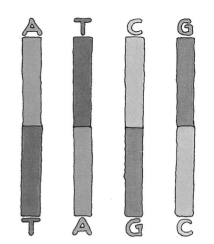

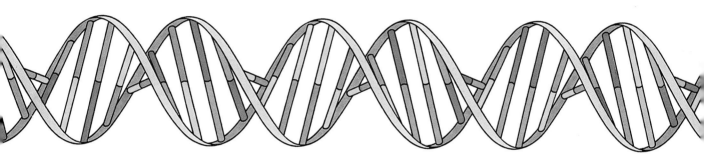

Can you see a pattern in the way the two DNA strands join together?

If you look carefully, you will see that chemical **A** always joins to **T** and chemical **C** (you will also see) always joins up with chemical **G**.

How do those four chemicals make DNA work inside your cells?

The first thing you should know is that DNA can copy itself. Inside the cell, the double helix of DNA splits apart.

Each single strand creates another strand from the chemicals—**A**, **T**, **C**, and **G**—that are already within the cell, matching **A** with **T** and **C** with **G**.

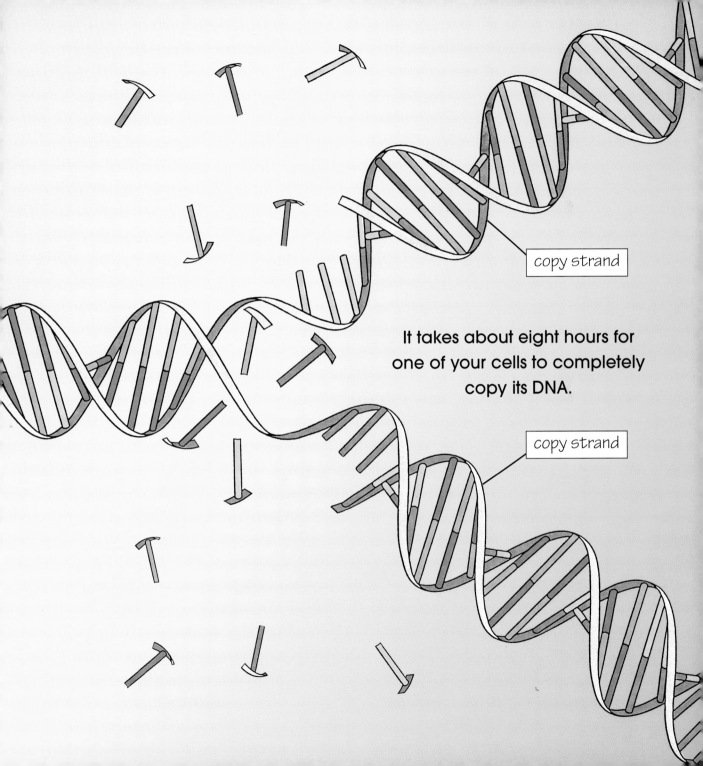

copy strand

It takes about eight hours for one of your cells to completely copy its DNA.

copy strand

What happens next in a cell with two copies of DNA? Well...

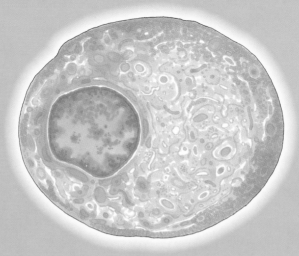

... the cell rests for a while (copying DNA is hard work!).

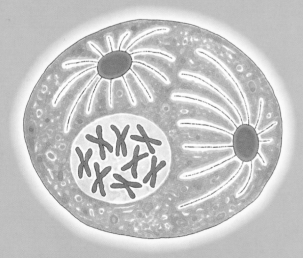

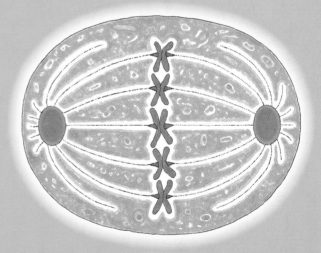

Then the DNA strands get shorter and fatter—you can see the individual chromosomes (remember page 6?).

The chromosomes line up in the center of the cell. They are attached to very fine cables.

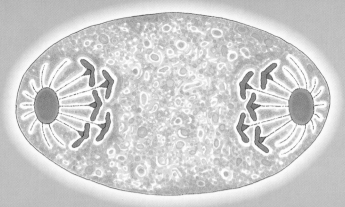

Each chromosome breaks into two halves with identical DNA. The two halves are slowly pulled apart by the cables.

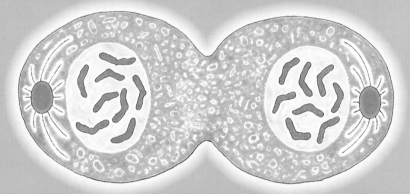

The cell begins to split in half. This will make two cells with identical DNA instructions.

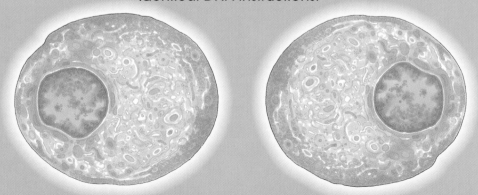

Millions of times a day inside your body, DNA is copied and cells divide. This helps you grow bigger and replaces old and useless cells.

The next thing to know is that DNA tells the cell how to make proteins. You could think of your DNA as a microscopic cookbook, filled with DNA recipes for many different proteins.

A DNA recipe for making proteins is called a gene. Here is how the cell reads each DNA recipe and makes proteins.

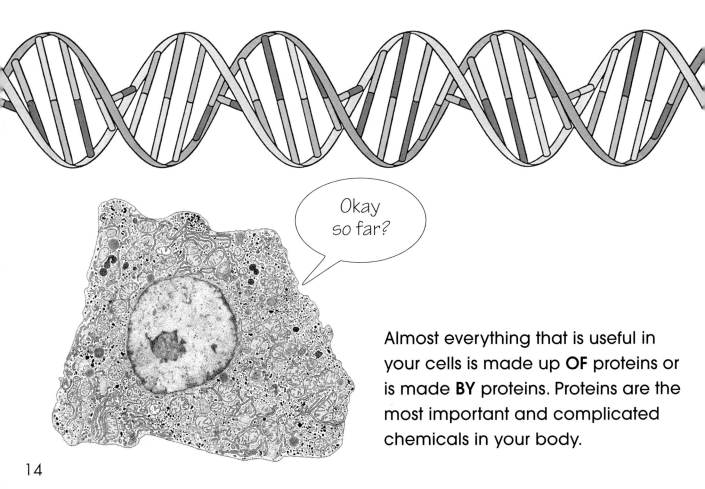

Okay so far?

Almost everything that is useful in your cells is made up **OF** proteins or is made **BY** proteins. Proteins are the most important and complicated chemicals in your body.

When your cells need to make a particular protein, the DNA gene recipe for that protein unravels a little bit.

A copy strand, called a messenger strand, is made. The messenger strand is made of **RNA**, which is short for ribonucleic acid (**R**ye-bow-**New**-cley-ick **A**cid).

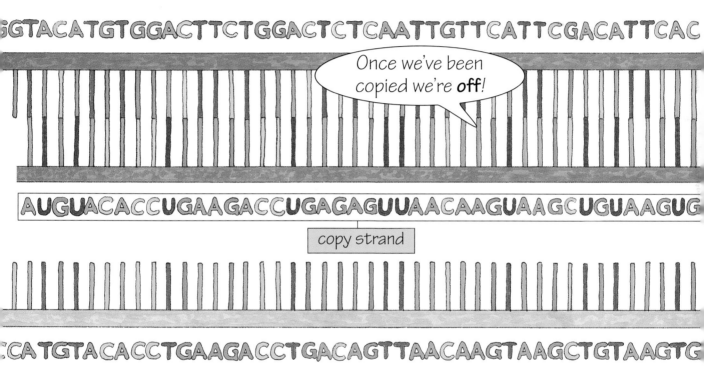

GGTACATGTGGACTTCTGGACTCTCAATTGTTCATTCGACATTCAC

*Once we've been copied we're **off**!*

AUGUACACCUGAAGACCUGAGAGUUAACAAGUAAGCUGUAAGUG

copy strand

CATGTACACCTGAAGACCTGACAGTTAACAAGTAAGCTGTAAGTG

RNA is slightly different from DNA. Instead of the **T** chemical, it uses a chemical called **U** for **U**racil.

So to make the single copy strand, **C** still joins to **G**, **G** joins to **C**, **T** joins to **A**, but **A** joins to **U** to make RNA.

The messenger strand then leaves the nucleus and goes into the cytoplasm. Here it meets a tiny micro-machine called a ribosome (rye-boh-sohm).

The ribosome "reads" the RNA, and makes a long string of amino acids. Amino acids are the building blocks of proteins.

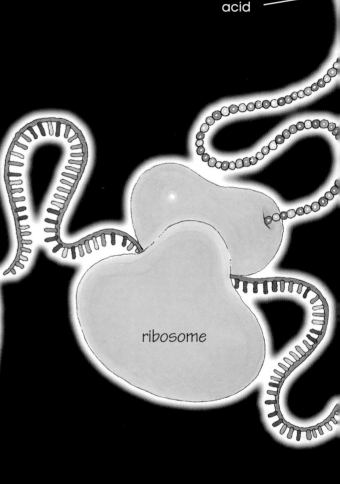

amino acid

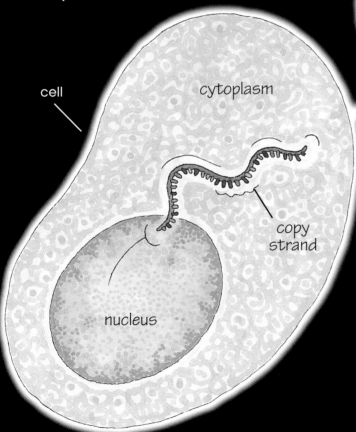

cell

cytoplasm

copy strand

ribosome

nucleus

They join together in a precise order like beads on a microscopic necklace. Then this "necklace" folds up into a complicated shape that is unique for each protein.

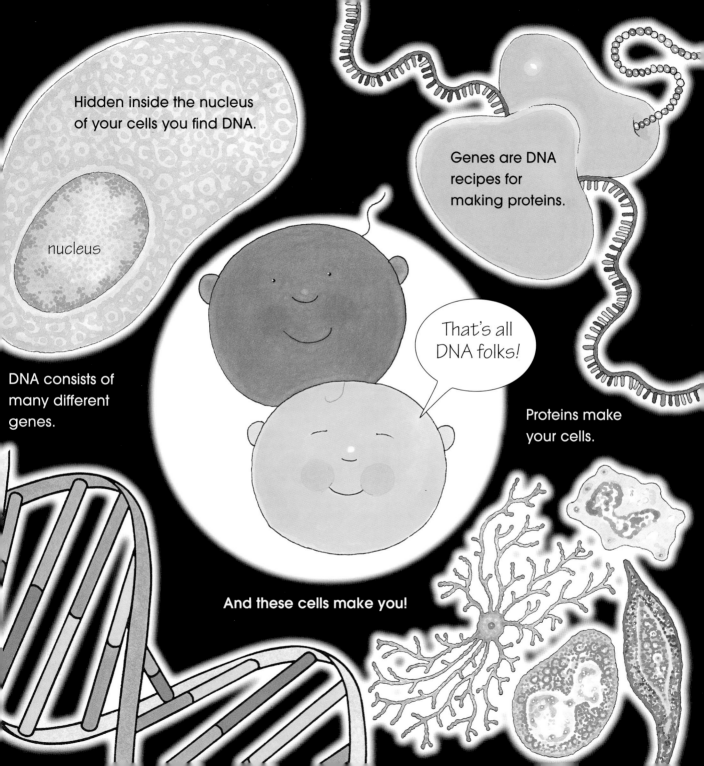

Your microscopic DNA contains an enormous amount of information! The human DNA code is made up of about three thousand million **A**, **T**, **C**, and **G**s on each side of the DNA strand. If you were to start reciting the order of the **ATCG**s in your DNA tomorrow morning, at a rate of 100 each minute, 57 years would pass before you reached the end (provided you did not stop to eat, drink, sleep, or go to the err...).

I'm bursting!

AT**C**G

Here is just a tiny bit of DNA that is the recipe for keratin (care-at-in)— your hair is made of keratin.

gagaatttag	actctgtctt	cagccaggca	ctccctccct
ccctcccagc	actatgccct	acaacttctg	cctgcccagc
ctgagctgcc	gcaccagctg	ctcctcccgg	ccctgcgtgc
cccccagctg	ccacagctgc	accctgcccg	gggcctgcaa
catccccgcc	aatgtgagca	actgcaactg	gttctgcgag
ggctccttca	atggtagcga	gaaggagact	atgcagttcc
tgaacgaccg	cctggccagc	tacctggaga	aagtgcgtca
gctggagcgg	gacaacgcgg	agctggagaa	cctcatccgg
gagcggtctc	agcagcagga	gcccttgctg	tgccccagtt
accagtccta	ttttaagacc	attgaggagc	tccagcagaa
gatcctgtgt	accaagtctg	agaatgccag	gcttgtggtg
cagatcgaca	acgccaagct	ggctgcggat	gatttcagaa
ccaagtacca	gaccgagctg	tccctgcggc	agctggtgga
gtcggacatc	aacggtctgc	gcaggatcct	ggatgagctg
accctgtgca	agtccgacct	ggaggcccag	gtggagtccc
tgaaggagga	gctgctctgc	ctcaagagca	accatgagca
ggaggtcaat	accctgcgct	gccagcttgg	agaccgcctc
aatgtggagg	tggatgctgc	tcccactgtg	gacctgaatc
gggtgctgaa	cgagaccagg	agtcagtatg	aggccctggt
ggaaaccaac	cgcagggaag	tggagcaatg	gttcaccacg
cagaccgagg	agctgaacaa	gcaggtggta	tccagctcag
agcagctgca	gtcctaccag	gcggagatca	tcgagctgag
acgcacagtc	aacgccctgg	agatcgagct	gcaggcccag
cacaacctgc	gagactctct	ggaaaacacg	ctgacagaga
gtgaggcccg	ctacagctcc	cagctgtccc	aggtgcagag
cctgatcacc	aacgtggagt	cccagctggc	ggagatccgc
agtgacctgg	agcggcagaa	ccaggagtac	caggtgctgc
tggatgtgcg	tgcccggctg	gagtgtgaga	tcaacacata
ccggagcctg	ctggagagcg	aggactgcaa	tctgcccagc
aatccctgtg	ccacgaccaa	cgcgtgcagc	aagcccatcg
gaccctgtct	ctccaatccc	tgtacctctt	gtgtccctcc
tgccccctgc	acaccctgtg	ccccacgccc	ccgctgtggg
ccctgcaatt	ccttcgtgcg	ctagaaccta	gggaatgcca
gaggagcaag	gatgcagggc	ccaggactcc	agagctgtga
cctggctctg	gttcaacaaa	aggggcctga	aaacatcatt
tgcatggctg	gagttgcccg	cgtaaggcag	ccaagaaact
cacccaaagc	ctgtagcctc	cccaactact	ccagactgtc
ctgctcaccc	tttccttcct	gggggtctgt	tccttcctat
gctcacccag	agaactctct	gatgtgccag	tgggcctccc
ttttaacctc	ctaataaata	tcatttcctt	ggcaaagcag
atg			

If you were to stretch out the DNA from those 46 chromosomes and lay it end to end, it would be over four feet in length.

Yet this DNA is coiled up inside the nucleus of a single cell.

And you could fit **one thousand** nuclei (new-clee-eye) across the period at the end of this sentence.

How can this be? Well, the DNA strands are incredibly thin. In fact, you could fit **ONE MILLION** threads of DNA across the period at the end of this sentence.

A little bit of DNA...

... sure goes a very, very long way!

Where did your DNA come from?

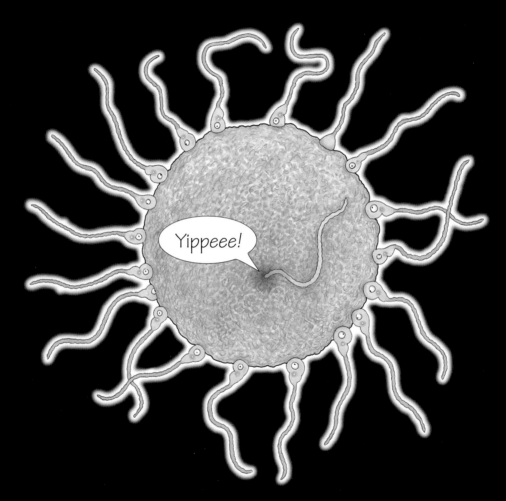

23 of your chromosomes came from a tiny cell called a sperm and 23 chromosomes from a larger cell, the egg.

They joined to make one single cell with the DNA code to make you, and only you.

The DNA in your first cell made a copy of itself. And then that one cell became two cells, each with a copy of your DNA. Those two cells became four cells, and four cells became eight cells, and so on until you were made of millions of cells. And each time a cell became two, a DNA copy was made.

That still goes on today. At this very moment, while you are reading this book, your DNA is working away. DNA and RNA copies are being made in millions and millions of cells inside your body.

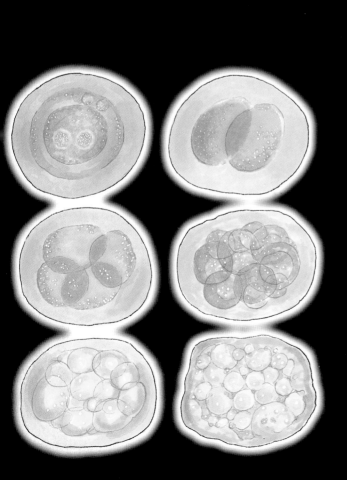

But your DNA is not just gene recipes for proteins. In between the genes, there are long boring random repeats of **C**, **G**, **A**, and **T**.

This DNA is not a recipe for any protein, and it cannot make an RNA copy. It does not seem to do very much, except copy itself.

97% of your DNA is pretty useless junk!

Now, it may not be of much use to you, but junk DNA has been extremely useful for detecting crimes. This is because the pattern of repeating bits in your DNA is unique, just like a fingerprint.

Scientists can extract DNA from the tiniest, invisible samples of skin, blood, saliva, etc., from the scene of a crime. They can measure the pattern of DNA as a unique "bar code." And if the DNA of the suspect matches that found at the scene…

It's a giant's fingerprint!

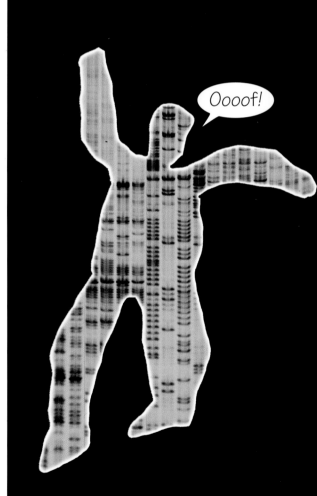

Oooof!

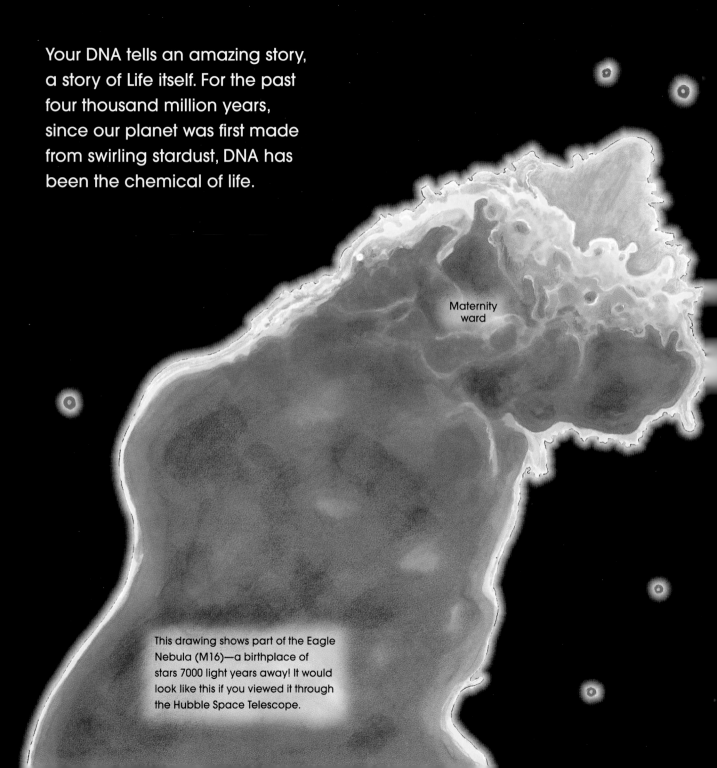

Your DNA tells an amazing story, a story of Life itself. For the past four thousand million years, since our planet was first made from swirling stardust, DNA has been the chemical of life.

Maternity ward

This drawing shows part of the Eagle Nebula (M16)—a birthplace of stars 7000 light years away! It would look like this if you viewed it through the Hubble Space Telescope.

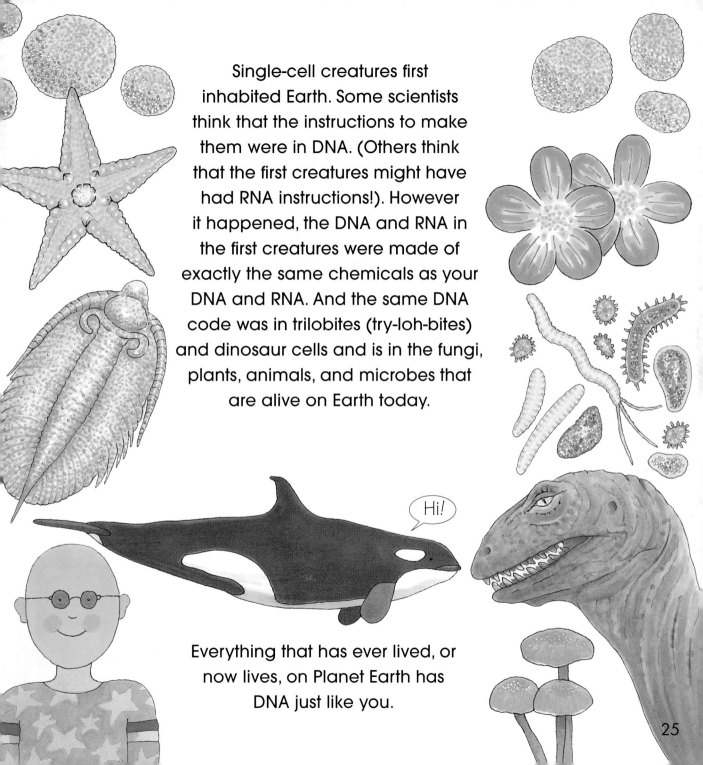

Single-cell creatures first inhabited Earth. Some scientists think that the instructions to make them were in DNA. (Others think that the first creatures might have had RNA instructions!). However it happened, the DNA and RNA in the first creatures were made of exactly the same chemicals as your DNA and RNA. And the same DNA code was in trilobites (try-loh-bites) and dinosaur cells and is in the fungi, plants, animals, and microbes that are alive on Earth today.

Hi!

Everything that has ever lived, or now lives, on Planet Earth has DNA just like you.

So why doesn't everything look the same? Why do you look different from a chipmunk or a woodpecker, a giant redwood tree or a mushroom, an armadillo or even a humble spider?

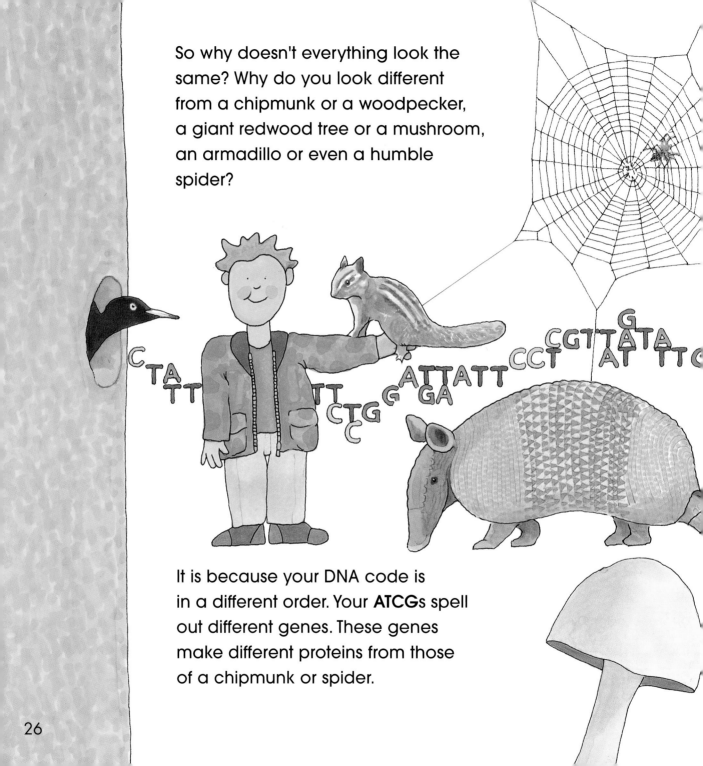

It is because your DNA code is in a different order. Your **ATCG**s spell out different genes. These genes make different proteins from those of a chipmunk or spider.

But a surprising number of your genes **are** similar to those of other creatures. Within your DNA, you can find echoes of the single-cell creatures that dominated our planet for its first three thousand million years; genes belonging to the animals and plants that evolved thereafter, and genes that were useful to your human and ape-like ancestors.

Some of your genes are similar to genes in flies and wiggly worms. You even have some genes in common with single-cell yeast that makes bread rise. In fact, over 30% of the known proteins in your cells are very similar to the proteins in yeast cells.

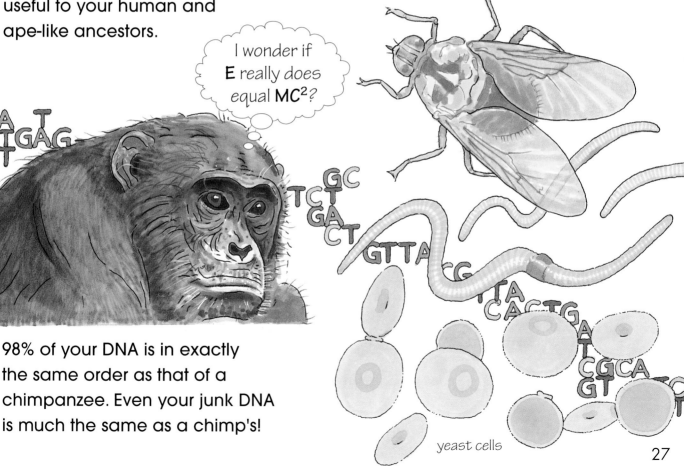

I wonder if **E** really does equal **MC²?**

98% of your DNA is in exactly the same order as that of a chimpanzee. Even your junk DNA is much the same as a chimp's!

yeast cells

27

The proteins are very similar, but they are not exactly the same. This is because of mutations (mew-tay-shons). Mutations happen when DNA is damaged. Mutations happen if mistakes are made as DNA is copied.

If the mutations are in the DNA of a gene, an altered protein may be made. Sometimes that altered protein may do slightly different, even better, work in the cell. This is why creatures are able to evolve.

Once again, look carefully at the DNA sequence.
Can you see where a mistake has occurred that will lead to a mutation?
Remember **A** should join with **T**, **T** with **A**, **C** with **G**, and **G** with **C**.

DNA mutations allow cells to change. DNA mutations allow cells to adapt to new surroundings. DNA mutations are the reason why there are millions of different species living on Earth today—and the reason why they are all related to single-cell creatures that lived four thousand million years ago.

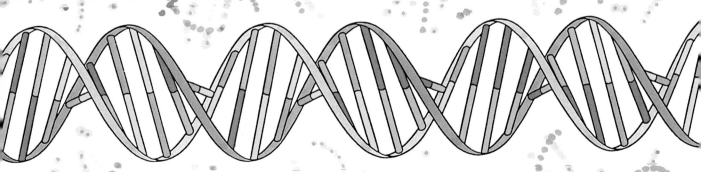

Using mind-boggling experiments and the world's most powerful computers, scientists have almost unlocked the hidden secrets of human DNA.

In the Human Genome Project, computers have "read" more than two and a half thousand million of the **ATCG**s that are found in human DNA. Now the scientists have to sort out junk from genes.

And they still do not know exactly how many genes it takes to make a human. All scientists can say is that there are at least 30,000 different genes in your DNA.

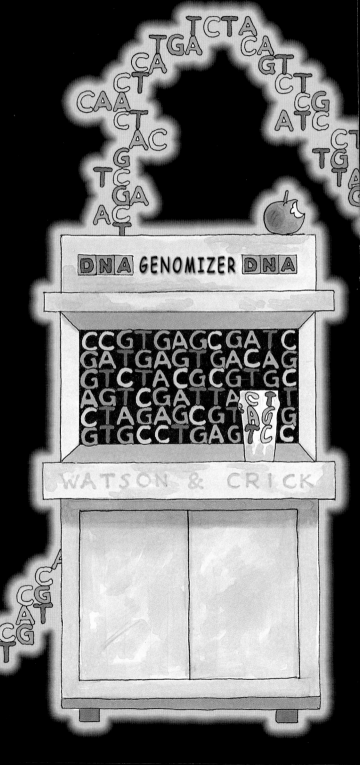

The Human Genome Project will touch the lives of every human being. Damaged genes and damaged proteins cause many diseases.

If humans can understand this, we can develop new and better drugs. We could also screen for damaged genes at birth so that the proteins can be repaired or replaced.

The greatest challenge of all will be to understand how your DNA and all your proteins work together to make one single cell. And how one hundred million million cells work together to make that incredible living creature

...**You**!